# SWEDENBORG HOUSE

44
———
200

# SWEDENBORG HOUSE

—

AND OTHER NUMBER COORDINATES

STEPHEN MCNEILLY

**HUNDTĒONTIĠ EDITIONS/ SWEDENBORG HOUSE PRESS**

2026

# 2

HUNDTĒONTIĠ

–

100


PRINTED IN A LIMITED EDITION Of 200 NUMBERED COPIES



PUBLISHED BY:
HUNDTĒONTIĠ EDITIONS/
SWEDENBORG HOUSE PRESS

DESIGN: THE QUINN FIZZLERS



ISBN: 978-0-85448-239-9

BRITISH LIBRARY CATALOGUING-IN-PUBLICATION DATA.
A CATALOGUE RECORD FOR THIS BOOK IS AVAILABLE
FROM THE BRITISH LIBRARY.

2026

*In heaven I have seen writing containing nothing but numbers arranged in patterns and series, and I have been told that such writing comes from the inmost heaven ... I have also been told that such numerical writing contains deep arcana, some of which cannot be grasped by thought or expressed in words. All numbers, in fact, correspond and have meaning depending on their correspondence, just as words do, but there is a difference: numbers represent general entities whereas words represent specific ones. And since one general entity involves countless specific ones, numeric writing holds more mysteries than the alphabet can ever contain.*

—Swedenborg, *Heaven & Hell*

CONTENTS

INTRODUCTION

# FROM 26 COLDBATH FIELDS TO 190 BLAKE AVENUE

STEPHEN MCNEILLY

Six of the images reproduced here were included in the exhibition *Concerning an idea about place*, staged at Swedenborg House between October 2022 and February 2023.[1] Drawn from a larger folder of miscellaneous paperwork, and with no visible markings except for a series of numbers scribbled by hand in pencil, the prints were bound in brown paper and tied together with string and a common reef knot. It took me some time to realize that the numbers written on the images were latitude and longitude coordinates, topographic reference points of houses and locations linked to the London cityscape of Emanuel Swedenborg and William Blake. The final three depart from Swedenborg House, heading eastwards to the London Hospital and then to Old Barking.

Since the Greek philosopher Pythagoras in the sixth century BC, numerous traditions have asserted a belief in the hidden power of numerical systems, Swedenborg included. One might think of G W Leibniz and the invention of infinitesimal calculus, Georg Cantor and the development of set theory and A N Whitehead and Bertrand Russell's *Principia Mathematica*. But it is only really since the development of the early computer model, introduced by Alan Turing, that a general expectation has emerged for number systems to provide a practical and essential foundation for the everyday and the commonplace. We find in numbers 'the ruler of all forms and ideas', Pythagoras is alleged to have said in a passage from his *Sacred Discourse*. In a similar way perhaps, at least according to Paul Dirac, might we now ascribe to numbers an inherent aesthetic or poetic value.

## INTRODUCTION

In a recent conversation with the writer Iain Sinclair, I mentioned once having woken from a dream to see multiple mathematical equations on my study wall, a series of numbers framing a coda of being 'in itself'. The numbers quickly dissolved however along with their promise. We then spoke animatedly of the possibility of a new poetic form, consisting solely of numbers, beautifully arranged and following an impulse divorced from the constraints of narrative and time, aligned instead with the more immediate relation between state and place. A double imprint. One related to internal mappings of our affections, and the other tracing external and lateral topographies, moving simultaneously east, west, north and south.

It is entirely possible, of course, via our binary-number systems and internet algorithms, that such poetic and correspondential networks already exist. To the untrained eye these systems might appear arbitrary or empty, serving primarily as conveyors of information between agentless hard drives. Yet from the standpoint of the angelic heavens, at least according to Swedenborg, we might in fact be glimpsing some ancient cosmic imprint, infinite in scope, in which are contained countless secrets, all lying poised until activated by a corresponding imagination.

It is in the spirit of this imagined, angelic sensibility, albeit more modestly, that the number coordinates are offered here. Drafted initially as pencil jottings in a notebook, they were then developed to mark the latitude and longitude coordinates between the designated locations, charting a trajectory through time and between states, a triangulation between image, place and number. Starting in 1772, they advance through to the present day and then retreat back to 1978. For anyone wishing to re-enact these pathways, a note is given at the end.

As always, it only remains now to thank those who have assisted with this project: to Iain for his inspired conversations; to Adam Skipper and James Wilson for their helpful comments; and to Terry McNeilly for his final words of consolation.

1 *Concerning an idea about place*, Swedenborg House Gallery, 13 October 2022 to 28 February 2023. An exhibition by Stephen McNeilly including works by William Blake, S T Coleridge, D T Suzuki, Emanuel Swedenborg, Vernon Watkins and others.

# SWEDENBORG HOUSE

AND OTHER NUMBER
COORDINATES

51, 523470
-0, 110454

## Coldbath Fields

51.523470 -0.110454

51.523370. -0.11052000. 51.524271. -0.10972320. 51.521370. -0.10622240. 51.522000. -0.10718000. 51.521655. -0.10561 080. 51.520986. -0.10501910. 51.520002. -0.10362220. 51.520151. -0.10245820. 51.520046. -0.10194890. 51.519130. -0.10199000. 51.518566. -0.10025160. 51.518160. -0.099380000. 51.517 216. -0.098346000. 51.516360. -0.098910000. 51.515833. -0.099412800. 51.514602. -0.099011200. 51.514721. -0.097846.

## Swedenborg Gardens

51.510062 -0.063362

51.514721. -0.097846. 51.514829. -0.097503700. 51.513742. -0.096794900. 51.513997. -0.095207000. 51.513740. -0.095699 900. 51.512514. -0.096903500. 51.512826. -0.095811. 51.512645. -0.095189. 51.512538. -0.094599. 51.512432. -0.094138. 51.5122 78. -0.093440. 51.512151. -0.092732. 51.511811. -0.091273. 51.511697. -0.090640. 51.511564. -0.089921. 51.511417. -0.089192. 51.511236. -0.088280. 51.511150. -0.087765. 51.51 0921. -0.086194. 51.510734. -0.084982. 51.510648. -0.084059. 51.510574. -0.083201. 51.510427. -0.082579. 51.509927. -0.081228. 51.509500. -0.08024. 51.509506. -0.079984. 51.509 540. -0.078492. 51.509473. -0.077087. 51.509540. -0.076089. 51.509800. -0.075134. 51.509606. -0.074244. 51.509286. -0.073 718. 51.508825. -0.073053. 51.508378. -0.071744. 51.508518. -0.069866. 51.508872. -0.068203. 51.509319. -0.066873. 51.509306. -0.066948. 51.509359. -0.066004. 51.509353. -0.064952. 51.50 9353. -0.063697. 51.509393. -0.063107. 51.510062. -0.063362.

Swedish Church, Princes Square
51.510062 -0.063362
The Fleet

51,514975.
-0,105237.
The south

## Poppins Court

51.514575 -0.105237

51.514575. -0.105237. 51.514306. -0.105228. 51.514292. -0.105 483. 51.514279. -0.105812. 51.514220. -0.106226. 51.514213. -0.107320. 51.514206. -0.107787. 51.514180. -0.108212. 51.514167. -0.108679. 51.514709. -0.108892. 51.515019. -0.108732. 51.515 257. -0.108700. 51.515389. -0.109157. 51.515159. -0.109512. 51.514751. -0.109630. 51.514398. -0.109555. 51.514177. -0.109 705. 51.513850. -0.111165. 51.513723. -0.111808. 51.513525. - 0.112383. 51.513345. -0.113091. 51.513104. -0.113499. 51.512897. -0.114078. 51.512830. -0.114464. 51.512684. -0.114979. 51.512 570. -0.115462 51.512456. -0.115816. 51.512296. -0.116224. 51.512183. -0.116578. 51.512063. -0.116921. 51.511996. -0.11 7297. 51.511829. -0.117887. 51.511749. -0.118198. 51.511568. -0.118595. 51.511408. -0.118895. 51.511375. -0.119014. 51.511275. -0.119335. 51.510589. -0.120411.

## Fountain Court

51.510589 -0.1200411

51.510746. -0.120648. 51.510579. -0.121056. 51.510411. -0.121523. 51.510 245. -0.121944. 51.510043. -0.122451. 51.509943. -0.122647. 51.509688. -0.123205 51.509471. -0.123639 51.509266. -0.124068 51.509022. -0.124565. 51.508820. -0.124961. 51.508558. -0.125439. 51.508169. -0.126174. 51.507 950. -0.126608 51.507758. -0.126960 51.507706. -0.127335 51.507594. -0.127823 51.507548. -0.128231 51.507561. -0.128615 51.507583. -0.128947. 51.507648. -0.129578. 51.507701. -0.130152. 51.507753. -0.130693. 51.507 798. -0.131034. 51.508050. -0.131310. 51.508294. -0.131474. 51.508534. -0.131723. 51.508763. -0.132018. 51.509080. -0.132308. 51.509352. -0.132509. 51.509598. -0.132775. 51.509841. -0.132992. 51.510022. -0.133196. 51.509 935. -0.133062. 51.510095. -0.133486. 51.510069. -0.134499. 51.510136. -0.134910. 51.510020. -0.135521. 51.509923. -0.136079. 51.509909. -0.136682. 51.509991. -0.137260. 51.510197. -0.137854. 51.510386. -0.138153. 51.510 726. -0.138470. 51.511002. -0.138823. 51.511378. -0.139163. 51.512234. -0.139930. 51.512662. -0.140332. 51.512510. -0.140187. 51.512856. -0.140491. 51.513200. -0.140793. 51.513579. -0.141121. 51.513919. -0.141439. 51.513 991. -0.141510. 51.513817. -0.142441. 51.513667. -0.143194. 51.513502. -0.144054. 51.513372. -0.144818. 51.513239. -0.145547. 51.513108. -0.146122. 51.513066. -0.146546. 51.513222. -0.146919. 51.513502. -0.147461. 51.513 612. -0.148011.

3 four Fan Court

51.510589 -0.120041

51.513612 -0.145011
17 South Molton street

## South Molton Street

51.513612 -0.148011

51.513562. -0.147585. 51.513342. -0.147108. 51.513230. -0.146865. 51.513020. -0.146305. 51.513133. -0.145957. 51.513194. -0.145664. 51.513268. -0.145340. 51.513303. -0.145155. 51.513356. -0.144890. 51.513416. -0.144608. 51.513467. -0.144305. 51.513598. -0.143531. 51.513671. -0.143133. 51.513956. -0.141675. 51.514024. -0.141396. 51.513752. -0.141144. 51.514072. -0.140038. 51.514350. -0.139046. 51.514839. -0.137206. 51.514027. -0.136639. 51.513503. -0.136271. 51.513296. -0.137758.

## Broad Street

51.513296 -0.137758

51.513775. -0.135197. 51.513921. -0.135043. 51.514077. -0.134533. 51.514433. -0.134719. 51.514652. -0.134877. 51.514819. -0.134978. 51.514923. -0.134593. 51.515037. -0.134176. 51.514966. -0.134081. 51.514841. -0.133925. 51.514977. -0.133358. 51.515132. -0.132801. 51.515220. -0.132472. 51.515293. -0.132203. 51.515409. -0.131761. 51.515545. -0.131336. 51.515654. -0.130951. 51.515771. -0.130477. 51.515825. -0.130220. 51.515871. -0.129847. 51.516016. -0.129424. 51.515969. -0.128875. 51.516040. -0.128599. 51.516193. -0.128144. 51.516314. -0.127791. 51.516493. -0.127376. 51.516580. -0.127236. 51.516516. -0.127163. 51.516664. -0.126703. 51.516719. -0.126529. 51.516888. -0.125992. 51.516972. -0.125790. 51.517044. -0.125478. 51.517088. -0.125068. 51.517161. -0.125097. 51.517223. -0.124977. 51.517337. -0.124769. 51.517504. -0.124347. 51.517732. -0.123813. 51.517803. -0.123632. 51.518016. -0.123043.

18 Broad Street
51.513296
-0.137758

51.51806 . -0.123043

## Swedenborg House

51.518016 -0.123043

51.518908. -0.121191. 51.517657. -0.120348. 51.517899. -0.117788. 51.518202. -0.115519. 51.518182. -0.111532. 51.517778. -0.107805. 51.517294. -0.104985. 51.516286. -0.102068. 51.515116. -0.097758. 51.514652. -0.096883. 51.514249. -0.094743. 51.513902. -0.092645. 51.513648. -0.090779. 51.513474. -0.089749. 51.513461. -0.088762. 51.513394. -0.086208. 51.513448. -0.084148. 51.513595. -0.081359. 51.513354. -0.079470. 51.513301. -0.077775. 51.513782. -0.076123. 51.514463. -0.073934. 51.515264. -0.071939. 51.516132. -0.069557. 51.516960. -0.067755. 51.517587. -0.066038. 51.518268. -0.063656. 51.518762. -0.061360. 51.518467. -0.062792. 51.518757. -0.061199. 51.518971. -0.060223. 51.518373. -0.059955.

## London Hospital

51.518373 -0.059955

51.517544. -0.059641. 51.517759. -0.057270. 51.516503. -0.057171. 51.514499. -0.057149. 51.514199. -0.053381. 51.513504. -0.04 8298. 51.512958. -0.041923. 51.512686. -0.038834. 51.512618. -0.036 380. 51.512549. -0.035920. 51.512413. -0.030925. 51.510920. -0.02 2052. 51.511029. -0.017100. 51.511247. -0.012280. 51.511384. -0.008337. 51.512311. -0.002509. 51.513074. 0.000339. 51.514465. 0.00 3494. 51.515409. 0.007667. 51.515089. 0.008237. 51.513998. 0.009300. 51.512696. 0.010384. 51.513487. 0.011633. 51.513644. 0.013670. 51.51 3725. 0.014908. 51.513603. 0.015708. 51.513678. 0.016000.

51,518373 -0,059465

London Map War

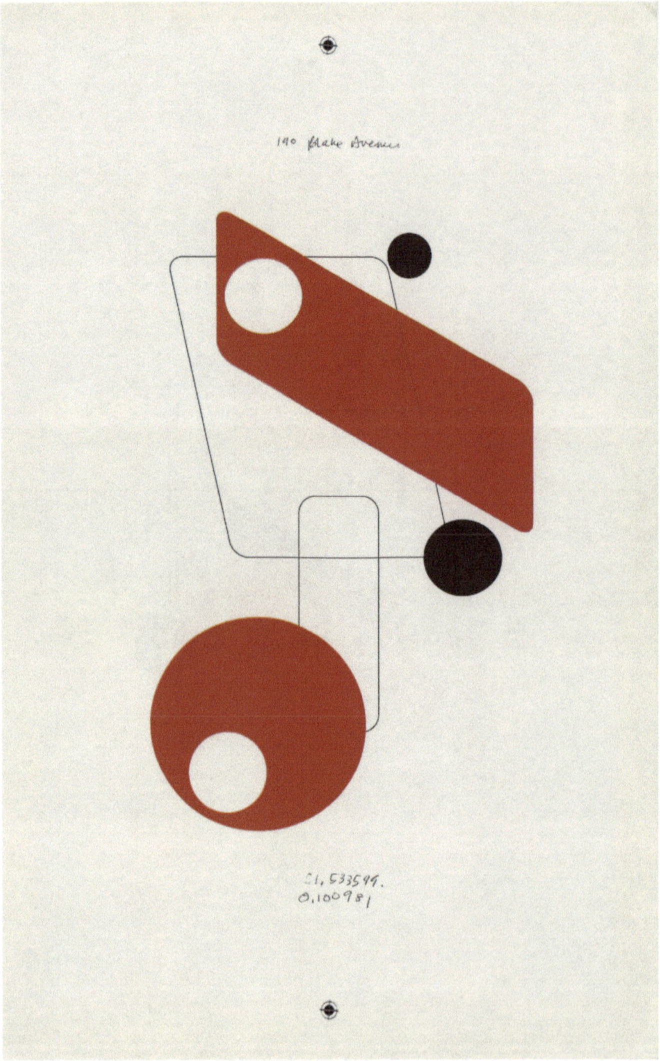
190 Blake Avenue
1,533599.
0,100981

## Blake Avenue

51.533599 0.100981

51.513678. 0.016000. 51.515169. 0.015830. 51.515362. 0.017150. 51.516 070. 0.017128. 51.516918. 0.017075. 51.517112. 0.017718. 51.517222. 0.020615. 51.517248. 0.023555. 51.516955. 0.028640. 51.517235. 0.033 232. 51.518610. 0.038103. 51.519625. 0.041193. 51.520092. 0.045377. 51.519732. 0.049776. 51.519452. 0.053638. 51.520199. 0.059067. 51.521 501. 0.062372. 51.523317. 0.067693. 51.525293. 0.073272. 51.527856. 0.080654. 51.529031. 0.086061. 51.530099. 0.092756. 51.532054. 0.091 361. 51.533389. 0.089601. 51.534644. 0.087456. 51.535204. 0.089687. 51.535565. 0.092842. 51.535590. 0.095014. 51.534992. 0.095197. 51.534 385. 0.095438. 51.533741. 0.095674. 51.533274. 0.095996. 51.532933. 0.096763. 51.532833. 0.097771. 51.532883. 0.098791. 51.532883. 0.099 864. 51.533377. 0.100652. 51.533599. 0.100981.

NOTES:
**Coldbath Fields**: the number sequence on page 13 marks the latitude and longitude coordinates between 26 Coldbath Fields and 1 Paternoster Row.
**Swedenborg Gardens**: the number sequence on page 14 marks the latitude and longitude coordinates between 1 Paternoster Row and the Swedish Church, in Prince's Square.
**Poppins Court**: the number sequence on page 17 marks the latitude and longitude coordinates between Poppins Court and 3 Fountain Court.
**Fountain Court**: the number sequence on page 18 marks the latitude and longitude coordinates between 3 Fountain Court and 17 South Molton Street.
**South Molton Street**: the number sequence on page 21 marks the latitude and longitude coordinates between 17 South Molton Street and 28 Broad Street.
**Broad Street**: the number sequence on page 22 marks the latitude and longitude coordinates between 28 Broad Street and Swedenborg House, Bloomsbury.
**Swedenborg House**: the number sequence on page 25 marks the latitude and longitude coordinates between, Bloomsbury and the London Hospital, Whitechapel.
**London Hospital**: the number sequence on page 26 marks the latitude and longitude coordinates between the London Hospital, Whitechapel and 56 Rogers Road, Canning Town.
**Blake Avenue**: the number sequence on page 29 marks the latitude and longitude coordinates between 56 Rogers Road and 190 Blake Avenue, Barking.

EDITIONS 100 HUNDTĔONTIĠ